THE STORY
OF A GIRL FROM
SIRIUS

AZARINE

Contents

Prologue

I looked out the large, curved windows of the place I called home, for now. Next to me was my friend, who would be known as "Ray" on Earth. My space name is Azarine. My Earth name, well, we will have to wait and see.

We had both undertaken to do Missions on Earth. To elevate Earth to a higher consciousness. Neither of us was thrilled to be living among revengeful, aggressive, warring Beings, but that was the whole point of the exercise. To uplift Earth.

Ray left a few minutes before I did.

On Earth, 40 years had passed. It would be decades until we found each other.

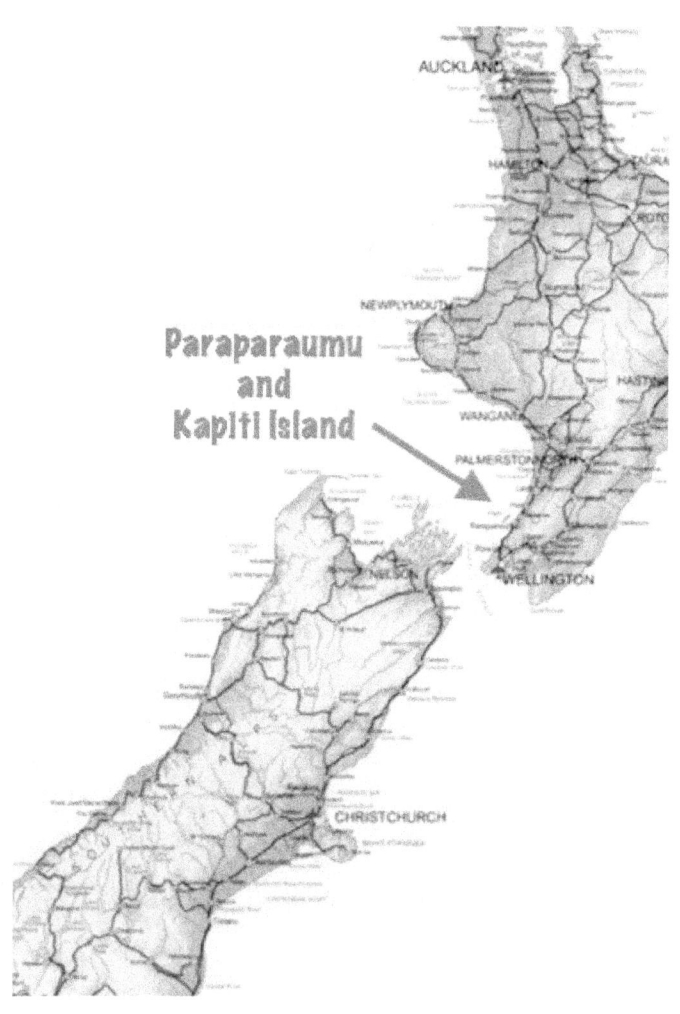

CHAPTER ONE
Was I Born, or Did I Walk-In to the Little Body?

I'm guessing I was born. There are memories of birth. Maybe, they are not my birth. Time will tell. My Mother, a Nurse and a Doctor are in the room as I open my eyes to a new world. I am popped into this family as the third daughter.

We go home. There is a new house. It is square. I am used to Hyperbolic Architecture. Curves and twirls and interesting designs. Earth is not like that. We live in boxes here. I have two older sisters who hold me. They are quite a lot older than me. Life moves on. I am breast fed, then dumped on a bottle. I am vaccinated. I do not respond well to vaccines.

Thankfully, we have cats. They are all Lyran. That's good, I have friends. I talk to the cats using telepathy. The cats purr back.

By three years of age, I am visited. Oddly enough, in the toilet, where I am alone. I swing my legs on the toilet, excited my galactic family have come, and look at the pretty dimensions they offer to me. The Beings who come are gnomes, fairies, and elves. I know who they are. They are my Space friends checking on me. I know exactly who they are, but I like them dressed up. They are so small. They are funny.

If you are unsure what a Walk-In is, read on...

There are tens of thousands of Walk-Ins on this planet. A Walk-In is a particularly high-minded entity who is permitted to take over the body of another human being who wishes to depart. A Walk-In must always have permission from the owner of the body.

There is a book called <u>Strangers Among Us</u> by Ruth Montgomery, that covers this topic.

CHAPTER TWO
Vaccines and School

Vaccines and more vaccines. This body does not like vaccines. I get sick. For years, I go on a merry-go-round of anti-biotics. As soon as I come off anti-biotics, I get sick. So back to the evil Doctor we go, and another round of anti-biotics is prescribed. I get sick. Every vaccine makes me sick. You'd think they would have figured it out. My lymphatic system is under stress. At night, I stare out the window looking at my home, 8.6 light years away.

My Mother wants to train to be a teacher. The day I turn five, she was out the door to teacher's college, but she brought home Cassius The Rabbit! He is my friend too. On the way to school, I get bullied by a boy called Christopher. No, not Christopher Robin. I wish I had Winnie the Pooh, Piglet, Rabbit and Kanga living with me. That would be fun.

As I progress through school, I figure a few things out. Firstly, I don't think Earth's math is correct. On Earth, we use the wrong base to count and calcu-late. Secondly, I think our science on Earth is completely wonky. It doesn't make sense to me. I keep my peace and play along.

I am seven years old. I awake one night and find an Astral Being at the end of my sister's bed. He is not doing anything to her. Just watching her. I believe he woke me, so I could understand that we all have an astral body. I watch him. He keeps his eyes on my sister, but I know he knows that I am awake and watching him. He looks ghostly, but I know he is not. He is a true teacher, a gift to me.

As I progress through Earth school, there are a few other visitations, however I am aware now that I have psychic abilities and can see auras, other dimensions and Beings who visit. As I enter high school, I am looking forward to getting out of it. There is little sense to what we are taught. It is not useful. After school, I work in a delicatessen.

We travel to Christchurch, my Mother and me to see some cousins. I end up in hospital where I have my appendix out. I can barely stand up straight after the operation, but I am determined to heal completely. The gash across my stomach will not be hidden by a bikini. I know this is vaccine-related.

By the time I escape school, I end up in University, studying a range of languages. Maybe I just want to reconnect with my own one. I clamber up Mac trucks and Kenworth trucks to mask them up for spray painting. I help a panel beater out, sometimes. I still work in the delicatessen.

Then, I quit university. I am not a quitter as such, but my parents have split up and I need to become independent – to have a job and earn some money. So that's what I do.

CHAPTER THREE
The Work Years,
The Little Entrepreneur

So, I bought a Café in a very popular place. It was before the invention of coffee machines. I work hard and make some money, then sell the place.

After that, I get a job working shifts. I'm not really different from any other young person on Earth. I work at a Camper Van place, helping people plan their trips around New Zealand, and showing them how a Camper works. With my love of languages, I am able to help out foreigners at the Camper Van place. We hire a Camper to the Rainbow Warrior bombers. We catch them as they (stupidly) returned the Camper after blowing up the Rainbow Warrior ship. I practiced my French with them to keep them busy. The Police came to arrest them.

A young man walks in to our office. He owns the big tour coaches that park in the yard. I immediately know that I will marry him. I take him upstairs to the Boss. Later, he asks if I will go out to dinner with him and some others, but I decline. I'm not that easy. He sends me cherries, telexes and cards from Wellington. He visits when he can. Finally, I meet him at his Lake Taupo house.

My Mother is on a teacher exchange in Australia for a year. I take care of the dogs, cats, bird, turtles and fish. I am an animal lover by nature. That's how I connect on this planet, with animals. A few days after my Mother returns to Auckland, I depart for Wellington with my cockatiel. That night, there is a massive earthquake in Wellington. It is the 21st January 1988. I won't ever forget it. I will give this man a year to prove he has good habits. He does. We get married on the 21st of January 1989. We have been married for 29 years, now.

In Wellington, I become a Real Estate salesperson, then, bored with that, I worked at the airport for a rental car company. It is a lively place to work. I worked shifts.

Then, we moved to Paraparaumu on the Kapiti Coast, and my life changes.

CHAPTER FOUR
Kapiti

gave birth to two girls. Neither are vaccinated, thanks to a couple I have become friends with. The man is a Naturopath and the woman works at a Natural Health Shop. We have extremely interesting conversations, involving various bio-frequency machines, herbs, and other topics. I become a raw fruitarian while in Kapiti, and have been for decades. We discuss the craft we see disappearing over Kapiti Island. Then, I find some research to support it, thanks to Bruce Cathie, a pilot at Air New Zealand, and Duncan Roads of Nexus Magazine.

My two lovely friends announce that they will get married shortly. We are invited. The wedding takes place locally and after the ceremony, I hear a name mentioned. I ask about the name and get pointed to an elderly man. His name is Ray. We talk and vaguely recognise each other. He is Horus from a past life in Egypt, but he is also my friend who looked out the windows with me. He says I was

Hathor in a past life. That fits very well with me. We find many people here, from our times in Egypt. Maat, Isis, Nephthys, Sekhmet, Anubis and others. I remember a lot of past lives here on Earth and other places. The most prominent is the one in a castle on the Isle Of Arran off the coast of Ireland. I remember the Vikings coming in their long boats to trade with us. Oddly enough, I have Viking blood in me this lifetime.

Ray and I spend some nights improving our psychic powers. We test food in the fridge, read the Urantia book and magically attract a lightship that visits, with the very people we have been reading about. We meet Ray's guide who astrally cloaks me for protection, and discuss so many topics my head spins in delight. The Urantia Book provides a comprehensive background to understanding the physical, mental and spiritual structure of the Universe.

I am studying to become a Naturopath too. I love plants and herbs and feel it is my calling. I pass everything and am top of the class. In Kapiti, I began a vegetable garden and a very large compost heap. The Pukekos come and steal the potatoes. However, I grew the biggest vegetables in the world. The cabbages, lettuces and broccoli were unbelievable!

At night I begin to go Astral. I learn to fly. I go to many places. I find out our children are also psychic. They know when our Bernese Mountain Dog is unwell. "Treacle doesn't feel well, Mummy." And with the little one, it is easy to play psychic tricks on her, popping colours up under her skirt so she lifts it to see what is there. I take them outside each night after going online to check the co-ordinates of the International Space Station. They watch it go over. I, however, am a little distracted by the other activity going on overhead.

One morning, half asleep I am visited by a man who asks if I will bear another child. I decline, saying two hands, two children. That is my lot in this world. He fades away.

I have other visitations too. Lots of them. Some good, some not so good. On my birthdays, I can hear my crew singing Happy Birthday to me, in the Language of Light.

My friends and I journey to local channellings in the area. I cough when the new Being arrives and have to leave the room. Personally, I cannot understand why they just don't appear as themselves and teach as themselves. We are all here to raise the frequency of Earth. We go to a lot of channellings. Every time, I get a cough and have to leave the room. I can see the owner of the body wandering around outside of his or her body. We make friends with some of the channellers, however I feel they are into a darker side of life and withdraw quickly. I only want Beings of Love in my life.

I begin to get enormous lumps under my arms. My first thought is cancer, from repeated vaccines. I juice fruit and they disappear. That's how you cure cancer. A juice diet.

CHAPTER FIVE
The Council

I can recall astrally travelling to some kind of Galactic Council. The room was magnificent, with a large table and ornate chairs. Everything was gold, or white and gold, slightly edged by blue. Not all the chairs had occupants. I took the chair that was mine and looked around at the faces. The faces are famous, Ascended Beings. They are beautiful, highly intelligent and beyond comprehension on Earth. We discuss matters related to the galaxy and try to solve them in a heart-centred manner. The meeting lasts well into Earth's night.

When it is time to leave, I return to Earth like a comet. I cannot describe it in any other way. It was like being catapulted back to Earth. And not that bad, actually. I got back into my body and patted my dog who knew I'd been away. He knows. I don't know how, but that dog knows. We are on Sirius.

I am not sure how regularly the Council meets but it is quite an experience. Some of the chairs do not have occupants and I gather from that the Beings are either away or, inhabiting a body elsewhere. At least they are only doing good for the galaxy. We all are.

CHAPTER SIX
The Gridlines

While in the library choosing picture books for my children, I noticed a book by someone I had heard of in a magazine. Captain Bruce Cathie, a pilot from Air New Zealand. I pulled the book from the shelf and sat down with my children. I was completely drawn into his world.

From my experiences, going astral at night, I knew about

gridlines. That's why so many ships were about. The gridlines cross not far from Kapiti. Maybe, there is a portal above Wellington. That would explain the number of craft there too.

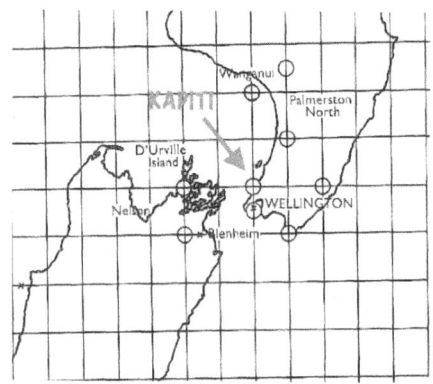

Regrettably, I never got to meet Bruce Cathie, however I do know some people who worked with him. They are Alec and

Brendon. Together they worked with Bruce, and Brendon created two PDFs that illustrate their findings. According to Bruce Cathie's theories, these lines are part of an electromagnetic/ geo-magnetic-ley/ scalar energy type grid for the use of these craft. They cannot be anything else. I have signalled them, messaged them, watched them, and know nothing here on Earth can match their speed and directional changes, not to mention their clever tricks when they know they have an audience of one. They disappear out of this dimension, then zoom back in. It's like watching live TV in your back yard. The grid appears to have been maintained over a very long, time period. I believe there could be Artificial Intelligence overlaid on top of it, but it doesn't stop the craft from getting to Earth. They are extremely advanced civilisations, and you will find out more later in this story.

CHAPTER SEVEN
Back to Wellington

The girls are in a Montessori School. Unfortunately, the options for Primary Schools are not that great, so we attend a meeting hosted by a private school and get hooked. We move to Wellington and have a bus stop outside our gate.

Things are good, we have a happy life, and I spend a lot of time at night outside in awe of the sight above me. There are so many craft, I am now convinced there is a portal around Wellington. Each morning, I get up early to meditate and find I have Beings who sit with me – well at least in the armchair I provide. They are fantastic to meditate with.

Our Bernese Mountain Dog died. I am wounded by it. Scarred by it. We go five years without a pet, but one day I pass a pet shop with the girls, and a little Siamese kitten comes home with us. They are being bred at cat farms, so really it was a rescue. The little Siamese is too small to get onto the bed so each night I tuck him under my arm. Now, eighteen years later, he still goes there.

Our children progress through school to high school – all at the same establishment. We go to Queenstown and Lake Taupo for holidays and live our lives quietly. I learn to trade the futures markets and sell on Amazon and learn other skills too. Several years later, Ray has moved, and we are no longer in contact, although he is a Facebook friend, but never posts anything.

CHAPTER EIGHT
2017

joined a CE-5 group (Close Encounters of the 5th Kind) which is human-initiated contact. I bought a tri-field meter,

an i-phone app and a laser pointer, which is technically illegal in New Zealand if you point it at an airplane, but not for a spaceship. It has been years since my team has been in contact. I miss them, but know they have other concerns around the galaxy. Sirians, Arcturians and Pleiadians and others, are around this planet taking care of us.

I still meditate each morning. One morning, I have my tri-field meter on, when a feeling of an unbelievable removal of all stress, comes over me. It washes down my body. This is what we are supposed to be like – no stress at all. A Lightship has arrived. The entire crew (well it feels like that to me) pours off the craft and the tri-field meter lights up and pings repeatedly. The first question I ask is, "Are we family?" using telep-athy. I get a resounding "yes!" in response. I am happy. They are back after such a long time. We spend quite some time together as they take in what I look like in a human body. It is amusing to me. My guides and my gang talk to each other and I feel a little left out. However, I join in asking other questions and get positive feedback.

Then, a few nights later I am on the craft. I know exactly where I am. I know where the entrance is to the big windows, I can see it. I suspect this is an updated ship and I am a little miffed that they dragged me into a side area of the craft. Perhaps the entire crew do not need to see me in my nightie. The purpose of the visit is to ensure I am on my Mission. I am. I try to explain that there are other things on Earth to be taken care of but get dragged back to the topic of the Mission. It's so hard not to fall in love with these Beings. They are heart centred, as am I. They are highly evolved, as I used to be. I can see their auras too, and they are all taller than me. Eventually, I fall back into my body in bed. After that, I find the craft every night and invite myself on board. I wander around. They don't seem to kick me off, so I watch what goes on and have no memory of it in the morning. I'm quite sure they wipe my memory before sending me back to bed. I wonder what I get up to on board, as I suspect some mischief goes on. I believe they have settled themselves above our house as there I more to the story.

One night, my husband is snoring like a trooper, so I ditch the bed and go up to my office. I put the trifield meter on and it pings. I am lying down, trying to sleep. Someone has realised I am not sitting up meditating and has come to check on me. How very sweet. Maybe I have a boyfriend on board!

They also appear in daylight hours. We had a CE-5 group meeting to attend on a Saturday, so I had invited my gang along too. They arrived outside our bedroom windows and my Siamese Cat alerted me by yowling loudly. He can easily spot them. I gave them the address and they disappeared. I have yet to beat them to a CE-5 group meeting. They always get there before me. Well, I guess it's easy in a craft and only takes a minute.

CHAPTER NINE
Harmonic Music

One fine day, I was looking for some Harmonic Music to listen to. I found a website called Sounds of Sirius and within seconds, my crown chakra blew open and several Beings were grouped around me.

They tried to operate my mouse, it was hilarious. I said, using telepathy, that I would do it. They know I loathe guided meditations, so we cruised away from those, and headed instead to a Language of Light CD. We played the sounds and apparently, they got a thumbs-up as I was told that was the CD for me. I downloaded the tracks and my gang disappeared. They are a very lively bunch of Beings. It was a very, very funny afternoon.

I have had that happen a lot when I am reading things online. All of a sudden there is a presence around me and everyone is reading my thoughts. I found other Harmonic music, and that got the thumbs up too. At least I am guided in my choices. I know they peer through the windows, watching life on Earth. It does not bother me, as how will they understand how we live if they cannot get a full appreciation of it. They need to witness it. They do come inside the house, but avoid the cat, as he screams at them. He is protecting me. My gang are peaceful and need to understand what people do, so they can fully grasp Earth, and the habits of its inhabitants.

CHAPTER TEN
The Secret Space Program

My prime concern is to get the Secret Space Program out to the world. It has been going on for almost 80 years and is known to some people around the world. Well, those who know about the tax extortion going on. Everyone is contributing their taxes to this project, but no one is getting benefit from it.

There are millions of humans out in the galaxy in spaceships and on bases on the Moon and other planets in our Solar System. It seems grossly unfair to humans, that so much technology is being hidden from them. They have the remedy for our environmental issues, world hunger and us no longer needing carbon-based fuel. We need those solutions, now!

They have replicators, anti-gravity craft, and so many other things it could cause a revolt on Earth if it is ever exposed. I would like to expose it. It is the 20 trillion dollar secret – repeated over and over.

You can find out
more online.

There are plenty of sites
disclosing it.

CHAPTER ELEVEN

Christmas Treats

While travelling from Wellington to Lake Taupo, I spot several craft hiding in clouds and message them. One over Taihape is a massive Arcturian craft so I invite them to our house at the Lake. It immediately comes closer. Then, crossing the Desert Road and going past the volcanoes, I see another Arcturian craft, just as massive over Mount Ruapehu. I invite that one too.

At Lake Taupo, we walk a lot. I go through the bush tracks (woods) on my own as often as I can. I like the peace and quiet. One day, I find a large blue sphere beyond some blackberries. I smile at the sphere, knowing exactly what it is. Spheres will carry you to craft, or to the Moon in under four minutes. I believe the Sphere presented itself as an alert to me for future use. Otherwise it would not have been beyond some prickly blackberries. I waved goodbye to it and carried on my walk, secretly delighted I had seen it. When I got home, little Arcturians were running around our deck. They are a bit shy, but I am hopeful I can befriend them. At night, there is activity overhead, but not as much as Kapiti or Wellington.

On our return to Wellington, there was a massive craft over Waiouru – in fact it was over the military facility there. I messaged it and received a message back that it was monitoring the base. Fair enough. I asked it if it could save the Kaimanawa Horses. I sent a mental picture of wild horses. Please save them, I asked, they will be killed soon. It's so unfair. I asked if they could gather more craft and transport them elsewhere, thereby upsetting the plans of the murderers. I could see smaller craft around the big craft, so I knew they were being protected.

CHAPTER TWELVE
The Craft in The Clouds

On our return to Wellington, I saw an enormous Mothership out a window of our home. I messaged it and it came closer to the point that it was almost overhead. The return message to me was a buzzing in my head, but sometimes languages simply don't cross over.

That night I went outside at about 3am to see if there was any action. I often do this, as craft do not always appear when humans are out and about – unless they are hiding in clouds. Anyway, a massive mothership was heading north, practically down our street. This thing was huge, enormous, and was not a meteorite. It disappeared out of sight and then went up out of Earth's atmosphere. There was other action as well. Craft zipping around way up high, but still easy to spot. I never fail to see something around here. We are not that far from Kapiti, so I look north too. The International Space Station always takes its time going over, however, real craft are very active.

One night I was blasted by the lights of a Triangle craft. Other craft blast me with light too. I'm guessing our place is a hot spot on the planet as they know someone friendly lives there. Little

Arcturians run past me and I try to instil confidence that I am one of them and will not harm them. I figure they know that, or they would not come at all. I tell them they are welcome to come inside but they are not ready yet. Time will bring them around. They are only about 3 feet tall, however I know of others who are 7 feet tall, but I have not yet encountered one, yet. The Arcturians are extremely advanced and have the most amazing craft. No computers. They don't need them. Mind power is everything. We have a lot of catching up to do, people!

Just as an aside, some of these craft I am signalling could easily be craft from the Secret Space Program. How would I know? But nevertheless, they signalled back so I can safely say I doubt they are, they are friendly folk who enjoy contact.

I'm still working on the wee Arcturians. They are so very precious. I love them and hope we will become friends.

CHAPTER THIRTEEN
Contact

Initially, our CE-5 group met in Karori, however we have people from Wanganui coming to join us so, instead we meet on Paraparaumu Beach. It's very close to the gridlines crossing. The amount of craft is simply unbelievable. I know there are more as I am messaging them, but they remain out of sight. We always have Beings drop in to visit us and join in with the guided meditations.

I always seem to pick up a lot more activity than others on the Beach, or in a room. My telepathy is improving, my psychic abilities are improving, and I feel that I am ready to go one step further and become a Breatharian. That will make life easier on the craft visits.

Sometimes there are months between visits, but I am assured that it is because of changes happening out in our galaxy. I do not want Earth or the Secret Space Program weaponising satellites or anything else. I also cannot believe that Elon Musk has not twigged yet that anti-gravity craft are the way ahead for us. Why is he still building rockets? I know his rockets return to Earth, but seriously Elon, go anti-gravity!

I know that the Secret Space Program has been accessing portals and sending massive amounts of freight through to Mars and other places, as it is so much easier using the portal system. Why do they not disclose that to the people of Earth? We too, need to join the galaxy. If we could get the word out to enough people, pressure could be applied. I will willingly be the spokesperson, however we need everyone on Earth to join together as citizens, put down the guns and weapons, and just come together as One.

I urge you all to join CE-5, just to learn the protocols. I mostly do everything alone these days, from going outside at night, to meeting my gang in my office.

Interestingly, no one has zipped down to check out this story yet. I have no doubt it will happen, but perhaps they are unaware of it. And just like that, as I type, someone is listening to the words in my brain. It's truly magical when you can connect with other advanced civilisations who are caring, kind and loving. They really do love us. There is no one else like us in the multiverse and we are quite unique with our 22 different ET genes and what not.

Yes, I did say that. No one on Earth is from Earth. We are all made up of ET strains and genetics. Unfortunately, one of the worst is the aggression gene. That makes for violence on the planet, and the religion gene keeps those who believe in religion, very occupied. The reason they keep us occupied is to keep us busy and not looking up.

But I look up, I know that they will likely project a hologram out across the planet making people think there is an alien invasion. The holograms are so amazing these days, it will look real, but it's not. Be warned. Think for yourselves. This is their last line of defence to instil fear. Don't fall for it.

Also, store food. There will likely be a solar flash that will melt everything – cars, computers, everything. You'll need food if that happens.

CHAPTER FOURTEEN
Alec

have a new friend. His name is Alec Newald. He is from New Zealand too. Maybe some of you know of him. He has written a book called CoEvolution. The book details his journey from Rotorua, a popular tourist town - to Auckland, New Zealand's largest city. The trip should only have taken Alec three hours at the most, but it actually took ten days. Alec drove through a cliff face and ended up on a spaceship. That spaceship took him across our galaxy to Haven, a planet that may once have orbited Earth. Alec spent time on Haven with Elders and his relations. They are his family, just like the Sirians, are mine.

Alec and I have had some amazing conversations, sharing our thoughts to each other, exchanging ideas that we both agree on, and our experiences on spaceships. Alec was brought back to Earth in ten days, but his memories were a little shady at first. Over time he has remembered so much, and his story is so endearing that I wish it had happened to me. It still might – but with my gang, not his.

I contacted Alec once, and asked if his crew were in his lounge. He could not tell at all. That night, I astrally flew there and took my gang too. We all telepathized with the Havenites and they accepted me too. The heart-centred galaxy around us, all get on. It's just the reptilian trouble on Earth and off-Earth that are causing strife. Alec slept through the whole party. What a shame, Alec!

Our conversations roam from Artificial Intelligence to the Grid where the craft arrive, to meeting new people and any ideas that come up that need discussing in a quiet manner to get to the heart of the matter under discussion. Alec shares interesting articles with me and I read and read and read them. Gaining knowledge is one of my favourite things, TV not so much.

We chat almost every day as we are so aligned on our thinking. One day, we will meet. Alec is a bit scared of me!

I don't beat about the bush and want to express my heart to people. I think that if we lived over the fence from one another we would never get anything done. I would be Space, Space and more Space all day long. And the nights would be spent with our spacey friends!

One thing that I found interesting in our daily chats, is that Alec's family were blue and so are the Sirians. Well, their auras are, and my aura is too, so I'm guessing several other civilisations could be too. The wee Arcturians are white when I see them around me. They are adorable, but I know they will trump me in the intelligence stakes. Their craft are totally amazing. I immediately knew exactly what I had seen when I saw the clouds. They came closer once I messaged them. They tend to do that once they know you are friendly. I invited them to our home as I know they need to understand how Earth works. They stayed on the deck, but I did invite them inside. PS: Note the nasty Chemtrail.

The Story Of A Girl from Sirius

The Ending

There is more to my story, but I did not want to bore you with Earth details. I'll update this one day.

Sadly, even as a young child, I knew my Earth family were not my true family. It was not always a good fit, and I understood that I needed to be in a family who would care for me, but I yearned for where I had come from. I learned to hide my abilities and gifts, as it was not appropriate to discuss them with my family.

People like me are called Blue Ray Beings. We have deep blue auras, and all come exclusively from Sirius. We have the Violet Fire in our auras and are powerful alchemists who are constantly transmuting energy, simply by our sheer presence in a place. We tend to spend a lot of time alone. Blue Ray Beings are the rarest on the planet. We are here to end the experiment, through purification and mutation. We are here to transform the damaged, mutated DNA of humanity. We are here for peace.

We are ultra-sensitive, empathic, intuitive, mystical and feel deeply. We are highly sensitive to foods, chemicals, the environment, noise and electricity. We are masters of Astral travel.

It will be easy for me to ascend. Blue Rays are more innately attuned to the spiritual plane than the rest of the planet. The Earth is moving to the 5th dimension.

Hang on to your hats!

No one on Earth,
is from Earth.

Every single one of you, is from somewhere else. A star system, a galaxy, or another Universe in the MultiUniverse.

I am from the stars,
not the swamp!

Author: Azarine
azarinefromsirius@gmail.com

Publisher: Earthshine Ltd